# How to Open Locks

# Without

# Keys or Picks

# How to Open Locks Without Keys or Picks

Many leading Locksmiths feel that opening locked doors by other means than picking or key fitting is a necessary part of a locksmith's knowledge. This point of view is correct, since the quickest method of entering is the best method from the standpoint of service and profit. A professional locksmith analyzes his job thoroughly before undertaking it. Thus he is able to determine before he starts what method to use. Since no two jobs present the same problem to a master locksmith, it is a credit to his ability for him to be familiar with all the ways of doing his work.

## The "Jimmy" and How to Use It

Best known among opening tools is the pry-bar, commonly called a "Jimmy." This tool may be a crowbar or a flat piece of steel varying in length from ten to twenty inches. A packing case hammer with a narrow steel handle and forked end is a small but extremely effective pry-bar. Plain flat chisels as well as heavy screw drivers have been used as pry-bars. In short, any tool that

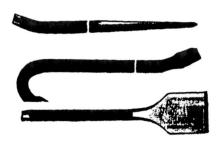

Fig. 1 Types of Jimmies

will pry, bend, or split a locked door or drawer is known as a "Jimmy." (Fig. 1)

At best the jimmy is a rough tool. Telltale marks almost always accompany every job. In cases where appearance is of little importance, the jimmy can be used quickly and effectively. If, however, no evidence of entry should be left, preparations must be made to prevent the damage or markings. Skilled workers are able to jimmy doors so that the bruises on the wood or metal are almost invisible.

The primary purpose of the jimmy is to pry the door and the jamb apart. To help this procedure, the stop or the strip of molding (A in Figure 2) along the outside of the jamb should be removed. This strip is usually tacked on with thin nails or secured by screws. By prying it loose or removing the screws in the vicinity of the lock, the crack of the door will be revealed.

A strip of wood (C in figure 2) should be placed along the jamb to absorb the pressure of the jimmy. By inserting the sharp end of the pry-bar into the crack near the lock the crack will be widened if pressure is used. A thin strip of metal (B in figure 2) should be placed on the edge of the door to save it from being splintered by the jimmy. A wedge of wood or metal (D in figure 2) should then be inserted in the crack to hold the door and jamb apart.

This process of prying and wedging the door should be continued only until the point where the type of locking bolt is visible. The next step depends on the bolt construction. There are several types of locking bolts, each of which is treated in a different manner with different tools.

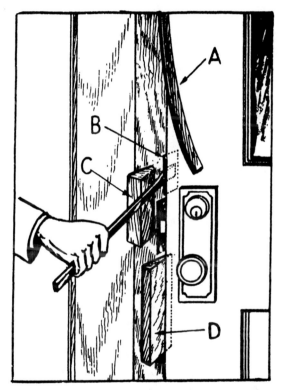

Fig. 2 Jimmyling a Door by Spreading and Wedging

# The Shove Knife and How to Use It

The ordinary springlatch has a bevelled bolt which is very easily pressed back into the lock by slipping an ordinary kitchen knife through the crack of the door. (Fig. 3) Notice A in the illustration. It is the molding strip that has been loosened from the door frame as described above.

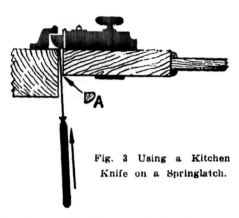

Fig. 3 Using a Kitchen
Knife on a Springlatch.

Most locksmiths avoid the unprofessional appearance of a kitchen knife by fashioning a shove knife that is notched as shown in Figure 4. Since some doors open toward the outside, while others open in the opposite direction it is necessary to have a set of 2 shove knives, with notches to correspond to the position of the bolt bevel. In the figure marked "pull" the shove knife is operated by inserting it over or under the bolt and drawing it back in a cutting motion along the bolt until it grips the bevel and forces it back.

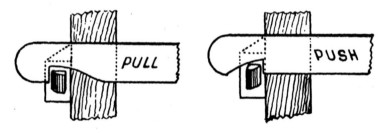

Fig. 4 Using Shove Knives to Trip Latchbolts

### USING CELLULOID AS A SHOVE KNIFE

A strip of celluloid or similar plastic may be used in the same fashion as the shove knife. It doesn't mar the wood or leave trace of entry. Moreover, in some cases where there is weather stripping or a solid jamb, it can be bent or twisted so that it can be inserted.

# Deadlocked Latches

If a latch is deadlocked by means of the small button on the case, the bolt cannot be tripped with a shove knife. Here is a case where the jimmy is used to force back the bolt with heavy pressure. It should be noted, that since most common latches can be deadlocked only from the inside, the locksmith should immediately consider the fact that someone may be inside. Jimmying, of course, is unnecessary if the person within can be aroused. It is well to be familiar with state laws before attempting jobs of this nature!

## Automatic Deadlocks

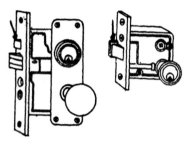

Fig. 5 Automatic Deadlocking Latchbolts

Some types of latches and mortise locks have mechanical arrangements that automatically deadlock the bolts when they engage the strikes. (Fig. 5) As in the case of the deadlocked springlatch, these types cannot be opened with shove knives. The jimmy is then used to force the bolts.

## Square or Dead Bolts

When a square bolt (Fig. 5) is encountered; a jimmy must be used. After the door and jamb have been spread and wedged, the point of the jimmy must be forced in front of the bolt. The object is to pry back

Fig. 6 Types of Square Deadbolts

the locked bolt and it may be expected that a part of the internal construction of the lock will break. Enough leverage should be used to accomplish the desired result. Keep prying until the bolt retracts. Short, jerking movements are better than one steady pull.

Lock sets that have both the latch bolt and the square bolt require two jimmying operations. However, the one spreading and wedging operation should be sufficient to reveal both bolts. In opening such locks, the square bolt should be jimmied first.

Fig. 7 Forcing the Deadbolt after Bending the Strike

If the latch were pushed back first it would only spring into the strike again.

In some cases the door cannot be spread far enough from the jamb so that the jimmy can reach the front of the square bolt. By pushing the strike forward, the front of the bolt may be reached. Most strikes, being manufactured out of thin plate, are easily bent as shown in Fig. 7.

If the front of the bolt cannot be reached, merely drive the point of the jimmy into the bolt so that it can get traction. (Fig. 8) Care should

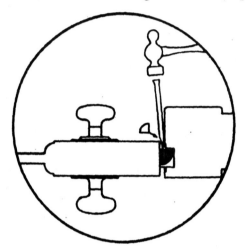

Fig. 8 Driving Jimmy Into Bolt for Traction

be taken not to raise burrs on the bolt. Such burrs may prevent the bolt from receding into the lock.

## Spreading Door Frames

A method of opening locked doors thot doesn't mar or splinter the frame or door is the Screw Jack System. Although this system requires somewhat heavier equipment, it is very efficient. A small jack

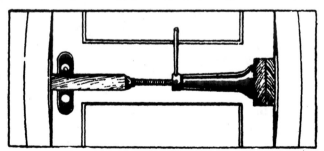

Fig. 9 Spreading the Door Frame with a Jack Screw

screw as shown in Fig. 9, a piece of 2" lumber about 28" long and several blocks of wood for spacers are required.

The blocks of wood are applied to the frame near the lock, as illustrated, being wedged into position by the long piece of wood and the jack screw. The blocks of wood are used as spacers for doors of different sizes. As pressure is applied by turning the screw, the frame gives way. Expansion of as much as two inches is not uncommon. Since the average lock bolt does not extend over three quarters of an inch, the frame is bellied out thereby freeing the bolt from the strike. The door can then be swung open.

## Jimmy-Proof Locks

Manufacturers, aware of the power of the jimmy, have developed many types of jimmy-resisting lock bolts. Several types are shown in Figure 10. Except for ripping the entire strike from the frame, the jimmy should not be used on these locks. Other methods of entering should be employed. The following paragraphs explain these other methods in detail.

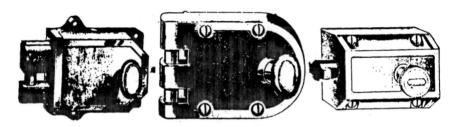

Fig. 10 Several Types of Jimmy-resistant Bolts

## Unscrewing Cylinders

A mortise cylinder is held in the lock case by means of a set screw or a cast iron fork, known to the trade as a yoke. In Figure 11 it will be seen that only the tip of the screw (or yoke, as well) engages the cylinder. By shearing off the tip, therefore, the cylinder would be free to turn. This is accomplished with a ten inch pipe wrench as shown in Figure 12. By gripping the rim of the cylinder in the jaws of the wrench, and turning hard in a counter-clockwise direction, the screw will shear. The cylinder can then be turned out. A paper washer should be made and placed over the cylinder to prevent scratching or marring the door or trim.

When the cylinder has been removed, the bolt can be drawn back by the fingers as shown in Figure 13.

Many locksmiths find it helpful to tear off the cylinder ring

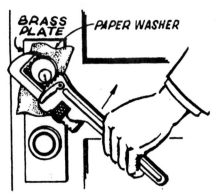

Fig. 11 Position of Set Screw
Against Cylinder

Fig. 12 Removing Cylinder with
a Wrench

before starting. This gives the wrench more gripping room. The ring is pried and torn off with a screwdriver and cutting pliers. New rings are inexpensive and can be installed when the job is finished.

A word of caution is in order: This method sometimes leads to the breaking of the lock case. It is especially not recommended for use on locks using the mortise Best removable core cylinder because an extra set screw is used in these cylinders to prevent successful entry by this method.

A                                    B

Fig. 13 (A) Action of Cylinder in Unlocking Bolt (B) Using the
Finger to Unlock Bolt When Cylinder Has Been Removed

# Twisting Rim Cylinders

An effective method of opening rim locks is to drive a heavy duty screwdriver into the keyway and to use a wrench to turn the whole cylinder. In so doing, the connecting bolts will shear, and the turning of the cylinder will operate the lock. Of course, in using this procedure, the cylinder has to be replaced with a new one.

# Drilling Cylinders

Cylinders may be unlocked by the drilling process. The purpose of drilling a cylinder is to destroy the upper pins or drivers that lock into the plug and prevent its turning. In other words, drilling destroys the shear line, and permits a screwdriver to be used to turn the freed plug.

The drilling should be done at the top shoulder of the plug, di-

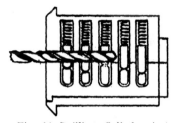

Fig. 14 Drilling Cylinder to Destroy Upper Pins

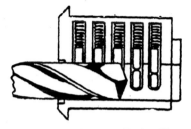

Fig. 15 Drilling Out Entire Plug

rectly over the key way. (Fig. 14) As each tumbler is pierced, the drill will jerk forward to bite the next one. After all tumblers are pierced, a screw driver can be inserted in the keyway and turned to open the lock.

Sometimes the remaining portions of the upper pins fall down into the plug after the drill is extracted, thereby locking it.

A lock pick is then used to raise each tumbler so that the locksmith can extract the small pieces or push them up out of the way to allow the plug to turn.

The Keil "Pick-Proof" Cylinder has a hardened steel plate that is difficult to drill through. Experience, however has proved that the plate is not as drill proof as it could be when high speed drills are used. In any case, however, the drilling may be done in the plug. As an alternative, the entire plug can be drilled out. (Fig. 15).

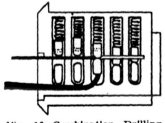

Fig. 16 Combination Drilling and Picking

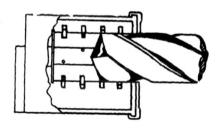

Fig. 17 Drilling Out The Plug of a Wafer Tumbler Cylinder

A combination picking and drilling method can be used effectively on pin tumbler cylinders. A one-thirtysecond inch drill is used with a sewing needle, and any lock pick. The hole is drilled into the cylinder about a thirty-second of an inch above the plug shoulder directly over the keyway. (Fig. 16) When all pins are pierced, a pick is inserted into the keyway. Each tumbler is raised while the needle is used to probe for the separating line between the upper and lower pin segments. After all pins have been separated, an ordinary screwdriver can be used to turn the plug. This method does not prevent the cylinder from being used again once the tiny drill hole is plugged up, and new upper pins are installed.

In the case of a wafer tumbler cylinder, the Ace or Illinois lock, a larger drill, about one-half inch in diameter is used. The object is to drill out the entire front part of the plug that contains the tumblers. When these are removed a screwdriver can be used to turn the remaining part of the plug. Care should be exercised to make certain that every tumbler has been drilled away. (Fig. 17).

## Opening Filing Cabinets

The drawers in inexpensive filing cabinets are locked by a vertical steel bar containing a series of hooks to engage the drawers and prevent them from being pulled open. A push-type pin tumbler cylinder is used. When pressed into the cabinet so that its face is flush with the

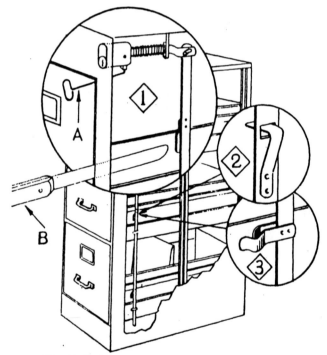

Fig. 18 Positions and Types of Locking Devices
in Filing Cabinets

front, the cylinder and bar is in a locked position. In opening these cabinets most mechanics use a long, thin piece of steel approximately one-sixteenth of an inch thick which is used to release the hook near the drawer which is to be opened. A long breadknife (B in Fig. 18) is ideal for this purpose. Because of different production designs, the hooks on different makes of cabinets are placed in different positions. Three types of hooks are shown in the illustration. By probing, the majority of these can be found and released.

Usually the hook is near the lock and almost directly behind it. By slipping the knife into the slit between the drawer and the frame, the hook usually can be located. An upward or downward pressure may release it. In some cases, the hook has to be pressed to the side. By probing with the knife, the Locksmith can recognize the construction. By experimenting he can soon determine the proper direction in which to press the blade.

In a few cases, the locking bar is placed in the rear of the cabinet. A short blade cannot reach it. When this occurs, the Locksmith will have to release the lock cylinder. The object is to reach the little spring plunger (at the end of A in the illustration) in the cylinder and

depress it. This operation will permit the cylinder to spring forward and release the locking bar. Probing, here, is also necessary because manufacturers have placed the latch in three different positions: top, side, or bottom. The plunger can be located by slipping a fine feeler gauge strip around the lock. (Fig. 19) Once the location of the plunger is found it should be depressed with a piece of hooked wire (Fig. 20) as far as it will go. A pick can be inserted into the keyway to pull it out.

Fig. 19 Feeler Gage Probing for Location of Spring Plunger

If a wire hook cannot reach the plunger, a one-eighth inch drill may be used to bore a hole in the top or side of the cabinet frame through which the plunger can be reached.

Fig. 20 Wire Hook to Depress Plunger

Some of the inexpensive wooden "Victory" or war model filing cabinets have escutcheon plates around the face of the lock. These are usually pried off with a screwdriver. Thereupon the plunger becomes visible and can be depressed with a pen knife blade.

Notice should be taken of any release button or lever near the handle of the drawer. Even though in an unlocked position, the drawer will not pull open unless this button is pressed. To try it would be similar to opening an unlocked door without turning the knob.

The better type filing cabinet is of fire resistant construction with overlapping edges that prevent the insertion of blades or jimmies between the drawers and frame. Unless an emergency requires jimmying, the cabinet should be opened by picking the lock or obtaining a key according to number, or by vibration. When confronted with such a cabinet, tilt it forward or sideward. Support the raised end with books, wood, or any other object at hand. Kick the supports away suddenly, and as the cabinet falls back into position, attempt to pull open the drawer. The vibration often springs the hook from the drawer and permits it to open.

## Opening Desks

The ordinary wooden desk is opened quickly when the Locksmith knows the construction of the two locking devices ordinarily employed. The Locksmith will encounter the remote control lock most often (4 in Fig. 21) In this construction, the center, top drawer contains the lock. When all the drawers are closed, the center drawer is locked, the whole desk is secured. The side drawers are locked by leverage action when the center drawer is closed tightly. The center drawer when pushed against the locking lever (4), forces the locking rod (3) down into the hook (6)

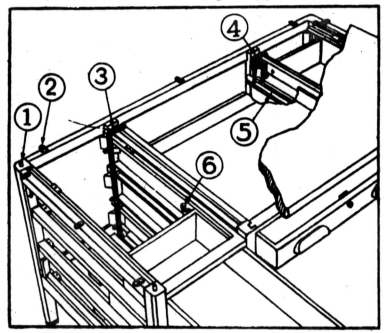

Fig. 21 Cutaway View of Rear Section of Flat Top Desk

on the end of each drawer. These desk drawers can be unlocked if the locking rod (3 in the illustration) can be reached and manipulated. These rods are usually underneath and to the rear of the side drawers. By reaching under, one can feel a wooden or steel rod. Raising this releases all of the drawers on one side. An extension of this rod can be found at the back end of the guide on either side of the center drawer. Just lift, or depress the extension, and one side of drawers will be free to open. A pair of pliers or a jimmy will be needed to remove or bend the plates that some manufacturers have installed to prevent access to these rods. Wooden panels can be removed by prying off the molding strips.

In opening a desk, the locking bolt should not be pried down. The object is to separate the bolt from the strike and to pull the drawer open.

Some flat desk tops are held in place by loose wooden dowel pegs (1 in Fig. 21) In order to release the lock of the center drawer, try to lift up the top first. If this doesn't yield, look for small angle brackets (2 in the illustration) that secure the top to the body. Remove the screws and lift the top free of the lock. Another method for releasing the center drawer is to unfasten the cross piece underneath the face of the drawer. An alternative to this is to remove the guide channels (5 in the illustration) of the drawer permitting the drawer to be lowered out of place.

Jimmying the top center drawer of a wooden desk is an easy matter. Most desk tops spring up. To avoid damaging marks, two strips of metal bent at right angles should be employed to protect the edges from bruises from the jimmy. In most cases, a mere quarter of an inch space is sufficient for the bolt to clear the strike.

The better type of metal desk does not yield as readily to jimmying. Chipped paint and bent rims usually result from using force unwisely. By covering the jimmy with a rubber strip from an old inner tube neater results can be obtained.

When spreading is not practical, the jimmy might be used to obtain enough space for a hack saw blade to be inserted. The bolt can then he sawed. This method is not recommended except as a last resort. Picking or obtaining a key by number are urged as better methods.

## Opening Letter Boxes

Careful jimmying is the accepted method of opening a locked letter box if the usual key fitting or picking is not practical. Because these boxes are made of thin guage metal, they give and bend readily. The locksmith should make a set of small jimmies to suit the delicate type of job required on a letter box.

Many locksmiths drill out the rivets or screws holding the locks. This requires a set of templates or guides which show the place and size of hole to be drilled when placed over the cylinder or keyway tube. Templates are made by the locksmith himself out of thin metal.

## Opening Windows

A window lock is a simple mechanism. When the upper and lower frames are parallel to each other, the lock can be opened by sliping a table knife between them and forcing the swivel to rotate back into the open position. (Fig. 22)

Weather resisting windows present a different problem because their frames are dove-tailed or beveled to shut out wind and rain. This arrangement does not permit entry of the knife blade. The approach in these cases is through the frame of the window. Since the woodwork on most windows is weather beaten and soft, it is not difficult to drill a small hole so that the swivel handle can be reached with a piece of stiff wire. An ordinary awl or even a nail bent in the shape of a crank can be used as a drill. The short length of wire should be bent into a slight arc so that when it engages the lock it can

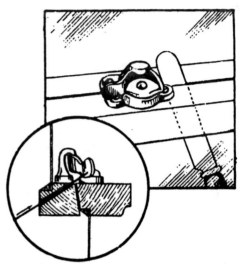

Fig. 22 (Square) Using Knife To Open Window Lock (Circle) Forcing Stiff Wire Through Drilled Hole to Push Open Swivel Handle

follow the path of rotation.

French or vertical windows are opened in the manner illustrated in Fig. 23. This type of window opens inward.

A glass cutter is a convenient tool for entering casement and cellar windows. A small hole, large enough for the arm, can be made near the lock. It is then simple to reach in to release the bolt. To avoid glass splinters caused by dropping the cut portion, a rubber suction cup or a piece of adhesive tape is attached to hold the loose piece once it is cut away from the pane.

Fig. 23 Opening French Window Lock

A diagonal line across the pane near a corner is the approved section of the window to cut.

## Entry Through Transoms

The lever and rod that controls the raising and lowering of the glass panel usually responds to strong pressure. By pressing hard on the transom window frame, one can push it ajar at least eight to ten inches. In most cases, this is the maximum distance because the bracket that supports the window will not extend further. The opening, however

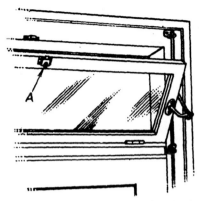

is sufficiently large to enable the locksmith to reach in with a screw driver. By removing the screws that fasten the bracket to the transom frame, the entire transom can be swung open.

Some transoms are fastened by an inexpensive spring latch. The blade of a bread knife inserted between the transom frame and the door frame will depress the latch, permitting the transom to drop away. A chain usually supports the transom in its open position. By clipping the chain, or removing the chain bracket screws the transom can be lowered its full distance.

Note:- Figure 24 shows the position where the spring latch is mounted. A spring latch is never used, in connection with a transom adjuster.

## Unlocking Barrel Bolts and Chain Door Fasteners

A barrel bolt (Fig. 25) is perhaps the strongest locking device used on a door. Because it operates exclusively from the inside, the only location where it can be attacked is at the crack of the door.

The ordinary barrel bolt slides into the strike. The small knob or handle drops into a notch provided to keep it in a locked position. Before the bolt can be retracted, this handle must be raised. The first step is to remove the weather strip. If a quarter of an inch space between the door and the frame does not appear, the locksmith will have to spread and wedge the frame apart for working space.

Fig. 25 Barrel Bolt

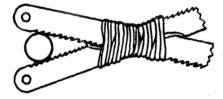

Fig. 36 Position of Tied Hacksaw Blades Around the Barrel Bolt

The next operation requires two hacksaw blades and a rubber band or cord. The blades are criss-crossed in scissor-like fashion, with the teeth pointing towards each other. The illustration (Fig. 26) shows the angle of the teeth. Using the blades in the same manner as a pair of pliers, the bolt is gripped and rotated, thus raising the handle from its notch. A side movement of the blades toward the unlocking direction will slide the bolt back a short distance. The sliding should be continued until the bolt clears the strike.

If the occasion should warrant, a keyhole hack saw or a single blade with tape around one end to protect the fingers can be used to saw the bolt in two.

A chain fastener is a device to permit a door to open from six to eight inches. Although the layman believes that it is a secure locking device, it is very easy to open. The equipment that is required is a strong rubber band or string, and a large thumb tack.

Another method of entry is taken from the old battering ram. Pounding the door with one's body will tear the bolt from its moorings.

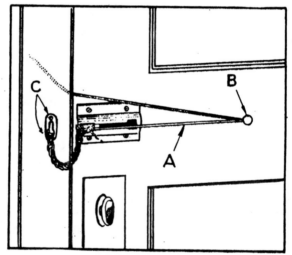

Fig. 27 Opening a Chain Fastener with a Rubber Band

When the door is in a closed position, the anchor of the chain is hooked into a track. As the door is opened, the chain is pulled to the opposite end where it is held in place under the pressure of the door. The chain cannot be retracted until the door is closed.

The object of the rubber band (A in Fig. 27) is to create tension toward the open end of the track. By slipping his arm through the door opening, the Locksmith can tie the band around the sliding anchor. He then extends the rubber as far as he can stretch it in a straight line toward the center of the door. The band should then be looped over the thumb tack (B in the illustration) which should be firmly imbedded in the door. A small string or thread tied to the tack and held by the locksmith will aid in retrieving it if it should fall out during the operation.

Upon closing the door, the chain will be drawn toward the releasing slot. The chain automatically travels along the track to the open end where it is in a position to drop out. If the anchor does not fall out of the track, it can be made to fall by shaking the door.

Another variation of the method of opening a chain fastener is

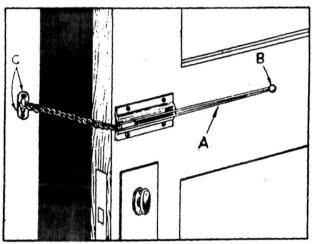

Fig. 28 Opening Chain Door Fastener With Cord

by using a piece of waxed string instead of a rubber band. (Fig. 28) The thumb tack is placed into position in the same manner as described above. The string is attached to the sliding anchor and looped around the tack. The loose end is held by the locksmith. When the door is closed, he merely pulls the string, causing the anchor to slide to the open end of the track.

If the door has a glass or steel panel so that a tack connot be used, he may unlock a chain fastener by removing the screws of the plate (C in the previous illustration) to which the chain is permanently attached. If the screws cannot be reached a pair of wire cutting pliers can be used to sever a closed link or bend an open link. An open link can always be bent back into shape and soldered.

## Opening Automobile Doors and Compartments

Automobile doors having ventilator panels in thir windows may be unlocked quickly and easily. The object is to open the ventilator so that the locksmith may reach the inside door handle. Once this is accomplished, the door may be opened by operating the inside handle in the conventional manner.

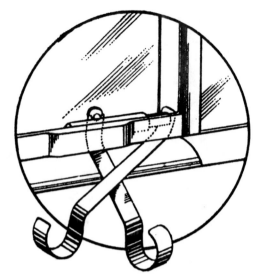

Fig. 29 Position of Metal Fingers When
Opening Slide Bolt on Ventilator

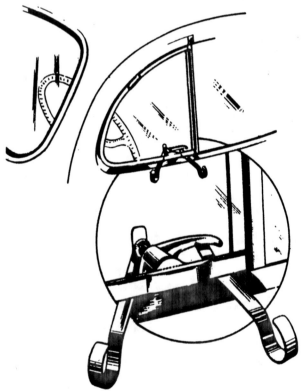

Fig. 30 Steel Fingers Opening Swivel Type Ventilator Locks

The first obstacle to opening the ventilator is the fastening device that secures the panel to the window frame. There are two popular types, the sliding bolt, and the swivel. The sliding bolt resembles the ordinary house bolt. The small knob, however, has a spring which has to be released before the bolt can be moved to the open position.

To release this bolt, two thin steel fingers are shoved underneath the frame of the panel. One of them is used to push the knob away from the frame; the other is used to push back the bolt. (Fig. 29)

In opening the swivel type, the same steel fingers are used. In this case, however, one finger is used to push the knob toward the frame, while the other is used to turn the swivel. (Fig. 30)

Releasing the fastener on the ventilator is only the first step in opening the door. The panel has to be opened sufficiently to permit the Locksmith to reach for the door handle. A screwdriver and a "link" are used to open the panel. (Fig. 31) The link is a flat, thin piece of steel with a hole in one end large enough to fit over the ventilator handle. The link is bent into a slight arc.

The screw driver is used to pry the panel sufficiently to allow the steel "link" to pass. through the crack between the ventilator and the main window. By hooking the "link" to the handle the locksmith can turn it until the ventilator opens. It is then a simple matter to reach in and open the door.

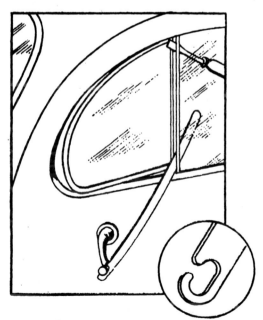

Fig. 31 Using Screwdriver and Notched Link
to Open Ventilator

Many auto door locks (notably General Motors) are secured by pressing a little button that is inserted in the window frame near the edge of the door. To open these doors a special hook was devised and made available to locksmiths. The illustration (Fig. 32 shows the tool and how it operates.

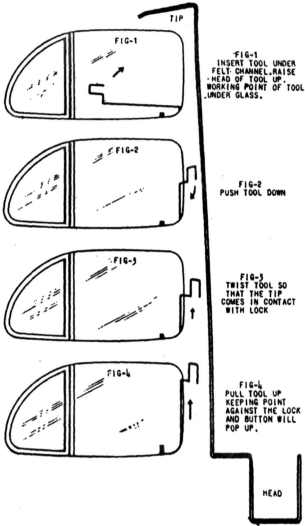

TIP

FIG-1
INSERT TOOL UNDER FELT CHANNEL, RAISE HEAD OF TOOL UP. WORKING POINT OF TOOL UNDER GLASS.

FIG-2
PUSH TOOL DOWN

FIG-3
TWIST TOOL SO THAT THE TIP COMES IN CONTACT WITH LOCK

FIG-4
PULL TOOL UP KEEPING POINT AGAINST THE LOCK AND BUTTON WILL POP UP.

HEAD

Fig. 32 Wire Tool to Open Push Button Type Lock

In Figure 1, the hook is inserted between the felt window channel and the glass. In Figure 2, the hook is pushed down about two thirds of its length. The next illustration shows how the hook is rotated so that the arm of the hook points into the lock. When slowly retracted, the arm anchors into the locking mechanism and raises it to an unlocked position. When the button pops up the lock is open.

Some of the later (1941-1942) car locks are difficult to "hook" because of a fabric covering over the lock. This fabric, however, can be torn off by the hook. Where a metal guard covers the lock as in the case of the more expensive cars, the hook will not operate. Entry through the ventilator panel is the better method.

Some cars do not have a ventilator panel or a button release on the lock. To open the door of such cars, the locksmith will have to remove the door handle on the passenger's (not driver's) side. This is accomplished by removing the escutcheon screws and by pulling and twisting the handle until it yields. By bracing one's knee and shoulder against the side of the door, enough pressure can be created to pull out the most stubborn handle.

To release the bolt, a screw driver is inserted through the handle hole in order to reach the rivets on the lock bolt. By bracing the screw driver against the rivets, the bolt can be pried back, thus releasing the door. (Fig. 33)

Fig. 33 Prying Open Car Door Bolt

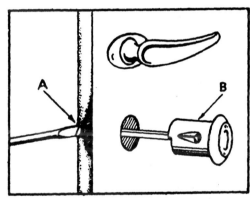

Fig. 34 Removing Door Lock Cylinder

On some Ford models, the door lock cylinder can be removed. This is true when the set screw (A in Fig. 34) that retains the cylinder is visible at the edge of the door. The cylinder and spindle should be removed. A long thin screwdriver inserted into the hub and turned will unlock the door so that the handle can be turned.

An old fashioned method for opening locked auto doors is to use a long piece of stiff wire, looped at one end. This wire is inserted through the clutch or brake pedal hole until it reaches the handle. After working the loop over the handle, the handle can be moved to the unlocked position.

# Ignition Locks

Ignition locks may be drilled out in the same manner as cylinder locks. As a last resort, Locksmiths have been known to chisel them out with small ⅛" chisels. These methods, however are crude. There is no question but that the usual lock picking or impression fitting systems are superior in these cases.

Automobile glove compartment doors are made of very light gauge steel. Because of this fact, they yield quite readily to a small jimmy. By using two screw drivers, both of which should be insulated with paper to prevent scratching, enough pressure can be exerted to spring out the door. It is advisable to pry down and then out. The underframe gives way while the door itself springs. No damage to the door or lock will result with careful handling.

Additional auto lock opening information will be found in other books written by the author.

## Tricks That Should Be Tried Before Using Tools

An analysis of a job often saves unnecessary time and effort. The quickest and easiest way to enter does not always require tools! Some of the obvious, though often ignored points of entry are discussed in the following paragraphs:

HIDING PLACES

Many persons are in the habit of leaving keys somewhere in the vicinity of the lock. Favorite hiding places are:

Tops of door frames
Under door mats
Under porch steps
In letter boxes
In milk bottles or containers

OPEN WINDOWS

Broken window locks, warped window frames, and sheer carelessness, leave many windows open. Before trying to enter any particular window, try several. One may be open already. This is often true of automobiles also.

OPEN DESK DRAWERS

Sometimes all the drawers in a desk are not closed securely. If one desk drawer is open, the entire drawer may be removed, permitting the locksmith to reach in and release the hooking device for all drawers.

## SPRUNG LOCKER DOORS

A sprung locker door indicates that entry had been made previously. Just pull hard. The door may spring open easily.

## FILING CABINET DRAWERS

Many filing cabinet drawers have been forced open because a key had been lost. These are rarely repaired and restored to their former degree of security. A quick pull may open them.

## PADLOCKS

A false locking action of the padlock shackle deceives many people. Upon closing the shackle, they hear a click and conclude that the lock is closed. A slight tap with a wooden handle of a hammer on the body of the lock will release the shackle.

## HINGE PINS

If the hinges are exposed on any door, panel, cover, or window it may be possible to punch out the hinge pins with a hammer and drift pin and remove the door. In certain cases, however, the pins are not exposed and have to be removed by other means. The average commercial and residential door, for instance, uses butt hinges with pins that can be lifted up and out with the fingers. There are other types of hinges where the tops and bottoms have been rivetted or pierced. These can be punched out after filing off the button head.

In this book consideration is given to the many methods for gaining entrance without picking or key fitting. This manual deals with the basic techniques as used by master locksmiths the world over. It should be explained, however, that the tricks are by no means limited. Personal cleverness and genius constantly develops newer and quicker techniques. Only standard methods are described. A locksmith with an inventive turn of mind will be able to find other methods on his own, and improve his skill in this art.